AIDS TO IDENTIFICATION OF
FLYING OBJECTS

Air Force Scientific Advisory Board

SAUCERIAN PUBLISHER

ISBNL 978-1-955087-08-7

© 2022, Saucerian Publisher

Al rights reserved. No part of this publication maybe reproduced, translate, store in a retrieval system, or transmitted in any form or by any means, electronic, mechanical, photocopying, recording or otherwise, without prior written permision from the publisher.

PROLOGUE

It is generally a good idea to return to the classics in any genre. This also goes for UFO literature. Rereading a book, or reviewing old documents after ten or twenty years is a rewarding experience. You will discover new data and ideas you didn´t notice before. The reason, of course, is that you are, in many ways, not the same person reading the book the second or third time. Hopefully you have advanced in knowledge, experience, intellectual and spiritual discernment. A good starting point is to reread the contactee classics material of the 1960s, in order to understand the deeper mystery involved in what happened during that era.

1947 is considered by most historians of the United States as the year when the Cold War began with the implementation of the "Truman Doctrine" to contain the propagation of world Communism, which give rise to the anti-communist hysteria of the following years. Also, 1947 was the year when the first UFO sightings were reported in the summer.

During the dawn of the UFO age, the United States government could not identify correctly those mysterious flying objects published a 40 pages booklet entitled: *Aids to identification of flying objects (1968)*. As a result that this saucer menace could be otherworldly beings or secret communist weapons, they decided to publish this booklet as a guide for properly identifying these flying objects.

This title has the following chapters:

INTRODUCTION
SHAPE
METEOROLOGICAL AND ASTRONOMICAL FACTORS
RADAR SIGHTINGS
PHYSIOLOGICAL ASPECTS
PSYCHOLOGICAL FACTORS
VISUAL PERCEPTION
QUESTIONS AND ANSWERS
BIBLIOGRAPHY

Saucerian Publisher was founded to promote Flying Saucer, Paranormal, and the Occult books. Our vision is to preserve the legacy of literary history by reprinting editions of books that have already been exhausted or are difficult to obtain. Our goal is to help readers, educators, and researchers by bringing back original publications that are difficult to find at a reasonable price while preserving the legacy of universal knowledge. This title is an authentic reproduction of the original US Airforce's *Aids to identification*. This book is a facsimile reproduction of the original printed text in shades of gray. Because this material is culturally important, we have made it available as part of our commitment to protect, preserve and promote knowledge in the world. This book has been formatted from their original version for publication. **IMPORTANT, although we have attempted to maintain the integrity of the issues accurately, the present reproduction could have missing and blurred pages poor pictures due to the age of the original scanned copy.** Because this material is culturally important, we have made it available as part of our commitment to protect, preserve and promote knowledge in the world.

                           Editor
                           Saucerian Publisher, 2022

EXCERPTS FROM A 1966 LETTER TO CONGRESSMAN L. MENDEL RIVERS, CHAIRMAN, HOUSE ARMED SERVICES COMMITTEE, FROM SECRETARY OF THE AIR FORCE HAROLD BROWN REGARDING UNIDENTIFIED FLYING OBJECTS

"... *Within the Department of Defense the Air Force has the responsibility of investigating reports on unidentified flying objects and of evaluating any possible threat to our national security that such objects might pose. In carrying out this responsibility let me assure you that the Air Force is both objective and thorough in its treatment of all reports of unusual aerial objects over the United States.*

"... *In order to evaluate this subject as thoroughly as possible, the capabilities of the Air Force Scientific Advisory Board have recently been focused upon the subject of UFOs. This Board ... completed a detailed review of this subject and concluded that the UFO phenomena presents no threat to the security of the United States, and that the present Air Force program dealing with UFO sightings has been well organized. Recommendations by the Board are presently under study and are expected to lend even stronger emphasis on the scientific aspects of investigating the sightings that warrant extensive analysis.*

"*Based upon 10,147 reported sightings from 1947 through 1965, (11,207 through 1966) ... I believe it is significant that the Air Force has succeeded in identifying 9,502 (10,532 through 1966) of these objects. Virtually all of these sightings were derived from subjective human observations and interpretations. The most common of these were astronomical sightings that included such things as bright stars, planets, comets and meteors, and fireballs and auroral streamers.*

"... *In evaluating these sightings, the Air Force has used carefully selected and highly qualified scientists, engineers, technicians, and consultants. These personnel have utilized the finest Air Force laboratories, test centers, scientific instrumentation, and technical equipment for this purpose.*

"*Although the past two decades of investigating unidentified flying objects have not identified any threat to our national security, or evidence that the unidentified objects represent developments or principles beyond present-day scientific knowledge, or any evidence of extraterrestrial vehicles, the Air Force will continue to investigate such phenomena with an open mind and with the finest technical equipment available.*"

FOREWORD

Unidentified Flying Objects (UFOs), or flying saucers as they are known to many persons, have been the subject of considerable interest on the part of the public as well as the U.S. Air Force for two decades.

On Dec. 30, 1947, the Air Force was given the responsibility by the Department of Defense for investigating UFO sightings.

This booklet is intended to meet the needs of individuals seeking information about UFOs. The Introduction is a brief history of UFO reports and studies and provides some information on scientific observations and analyses. The section "Aids to Identification of Flying Objects" is more technical and should be of interest to persons desiring in-depth information on flying objects and natural phenomena. The "Questions and Answers" section will help answer many of the questions commonly asked about UFOs.

TABLE OF CONTENTS

INTRODUCTION 4
   AIR FORCE RESPONSIBILITY 5
AIDS TO IDENTIFICATION OF FLYING OBJECTS ....
   SHAPE 10
   METEOROLOGICAL AND ASTRONOMICAL FACTORS 11
     Mirages
     Refractional Dispersion 12
     Planets 13
     Comets
     Meteors 14
     Sundogs and Moondogs 15
     Auroras 16
   RADAR SIGHTINGS 17
   PHYSIOLOGICAL ASPECTS 20
   PSYCHOLOGICAL FACTORS 21
   VISUAL PERCEPTION 24
     Sky Search
     Depth Perception
     Accommodation
     Seeing at Night

QUESTIONS AND ANSWERS 28

BIBLIOGRAPHY 35

INTRODUCTION

UFOs do not constitute a new phenomenon.

In 1254 at Saint Albans Abbey, England, an "elegantly-shaped, well-equipped ship of marvelous color" appeared in the sky.

A farmer in Texas reported seeing a "dark flying object in the shape of a disk cruising in the sky at a wonderful speed" in 1874.

Reported sightings were somewhat sparse until Kenneth Arnold made world headlines in June 1947 when he described a chain of nine fast-flying objects appearing ". . . like saucers..." near Mt. Rainier, Wash.

Scientists examined all the facts presented by Arnold. They determined that weather conditions at the time of the sighting were very stable and were likely responsible for an increase in the index of refraction* of the atmosphere (Arnold had reported the air was "clear as crystal").

After weighing all the available facts, the Air Force concluded that Mr. Arnold had witnessed a mirage created by uncommonly stable weather conditions.

During World War II, the strange illusions that sometimes occur during unusual atmospheric conditions presented a problem for members of a U. S. Navy task force in the Pacific.

Navy spotters reported a strange aerial object rapidly approaching their ship, and opened fire. The ship's navigation officer who was not on the bridge at the time, heard the guns and quickly returned only to discover the gunners were shooting at the planet Venus.

Atmospheric conditions had apparently made the gunners victims of an illusion of proximity.

*An index of the degree of bending of light rays following Snell's Law as this light or reflection passes from one layer of atmospheric density to another.

AIR FORCE RESPONSIBILITY

On Dec. 30, 1947, the Air Force was given the responsibility by the Department of Defense for investigation of UFO sightings under the code name "Project Sign." Air Force interest in the investigations, then as now, related directly to its responsibility for air defense of the United States.

The Air Technical Intelligence Center (ATIC) conducted the investigations.

"Project Sign" conclusions stated in February 1949 hold true to this day:

". . . It is unlikely that positive proof of their (UFO) existence will be obtained without examination of the remains of crashed objects. Proof of non-existence is equally impossible to obtain unless a reasonable and convincing explanation is determined for each incident. Many sightings by qualified and apparently reliable witnesses have been reported. However, each incident has unsatisfactory features, such as shortness of time under observation, distance from observer, vagueness of description or photographs, inconsistencies between individual observers, and lack of descriptive data, that prevents definite conclusions being drawn. . . ."

The UFO project continued on a reduced scale, and in December 1951 the Air Force entered into a contract with an industrial firm for a detailed study of UFO cases on file. It took the company three years to complete the study, which was released under ATIC cover because the company desired to remain anonymous in this field of research.

This report, commonly referred to as "Special Report #14," evaluated all UFO data in Air Force files.

The report stated: "It can never be absolutely proven that 'Flying Saucers' do not exist. This would be true if the data obtained were to include complete scientific measurements of the attributes of each sighting, as well as complete and detailed descriptions of the objects sighted. . . .

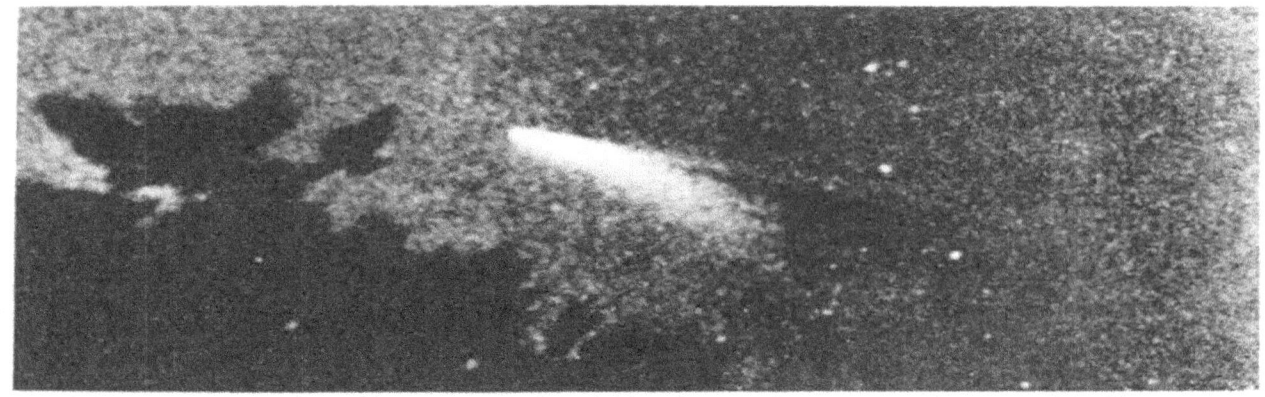

Photo of 1957 Comet, Mrkos, taken by Curtiss A. Griffin, helps explain why such phenomena are often mistaken for "UFOs" or "flying saucers."

". . . on the basis of this evaluation of the information, it is considered to be highly improbable that any of the reports of unidentified aerial objects examined in this study represent observations of technological developments outside the range of present-day scientific knowledge."

In 1949, "Project Sign" was changed to "Project Grudge," and in 1952 the program of investigating UFOs was given its present name, "Project Blue Book."

Of the 11,207 UFO sightings reported through 1966, all but 675 had been identified. Many of the unidentified remain in that category simply because sufficient information is not available to complete the studies.

During the 20 years of investigations and analyses, photographs have been submitted for evaluation in conjunction with UFO reports. The objects in these photographs have been determined to be mostly misinterpretations of known natural phenomena or conventional objects. Some have been determined to be double exposures, "trick" photography, hoaxes and flaws in negatives or in developing.

Services of qualified scientists and technicians have been, and will continue to be, used to investigate and analyze reports of unusual aerial phenomena over the United States.

In May 1966 the Air Force took steps to strengthen scientific investigations of UFO reports. The help of more individuals within the scientific community was solicited through contracts calling for prompt, in-depth investigation of selected UFO reports.

In October 1966 the University of Colorado was selected to undertake the investigations under an 18-month contract. Dr. Edward U. Condon, internationally known physicist, now a professor at the University, was named to head the study.

AIDS TO IDENTIFICATION OF FLYING OBJECTS

During the past few years, many flying objects reported as UFOs were later found to be manmade. Balloons and aircraft have often been misidentified.

For example, an observer who sees a silvery, transparent, disk-like object may be looking at a balloon. The absence of exhaust or engine noise, or any visible means of propulsion, would support such identification. Weather balloons are sometimes released in clusters and may drift in what appears to be a formation, depending on the air currents. They shimmer in reflected sunlight or moonlight, and seem to hover as they pass from one air current to another.

Upper air research balloons may attain great heights and travel great distances before they fall back to earth. They may be observed, therefore, in areas far removed from a launching site. Research balloons are usually constructed of material with a highly reflective surface. They are often 200 feet in diameter and are visible, under certain atmospheric conditions, at extreme heights. Such balloons, seen in reflected light, may seem disk-like in shape and may appear to have an oscillating motion. (They carry radar reflectors which can result in electronic contact.)

An observed object is not usually a balloon if the speed is very fast. However, some balloons travel in the upper air currents at speeds exceeding 100 m.p.h. To help identify a flying object as a balloon, the observer should keep in mind that a balloon moves with the wind and not against it.

To learn more about atmospheric pressure, temperature, wind direction and velocity over vast stretches of open sea, balloons are being launched to fly at 30,000 feet. They are rigged to destroy themselves if they drop below 28,000 feet, or fail to reach that height. These balloons are 40 feet in diameter and have a plastic skin only 2/1000ths of an inch thick. Sunlight or moonlight reflecting from their plastic skin surface can cause them to be easily mistaken for UFOs.

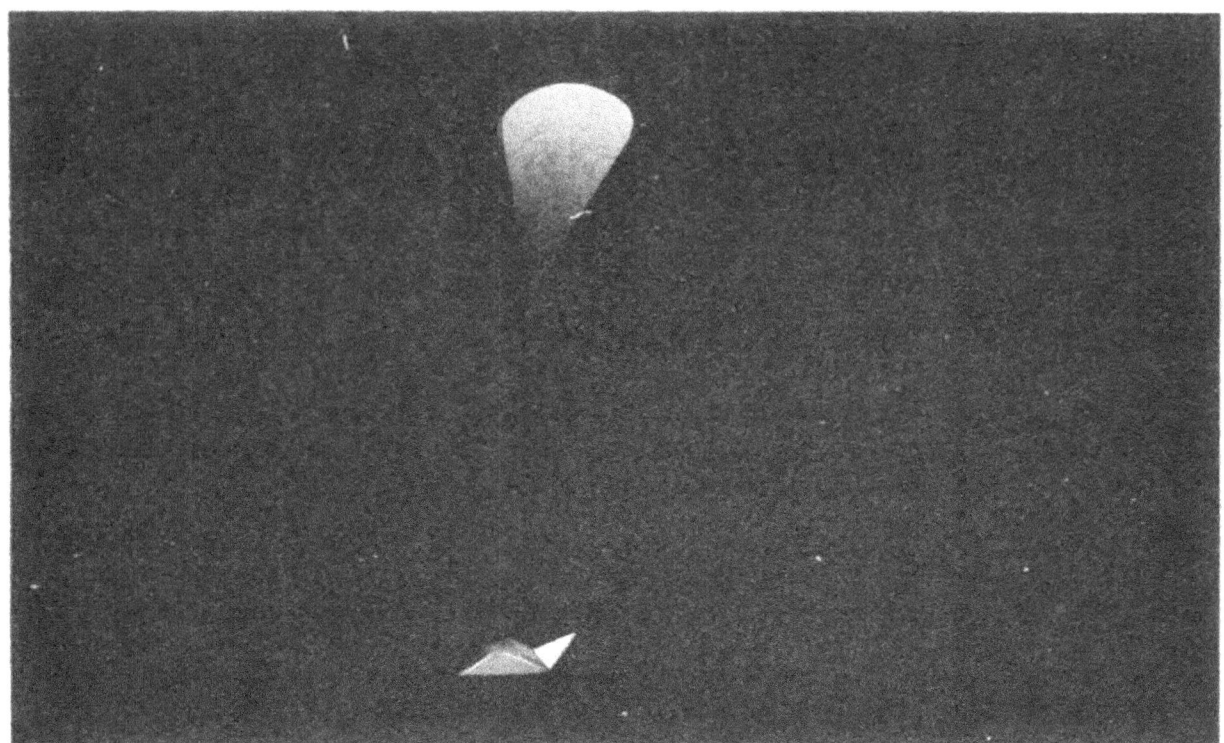

FIGURE 1: What may appear to be a UFO soaring through the sky may actually be an object such as this weather research balloon.

FIGURE 2: Radar planes may easily be mistaken for UFOs under unusual weather conditions.

Because balloons can be observed when the sun is below the horizon and the earth is in comparative darkness, a considerable number of them have been reported as UFOs over the past 10 years. These UFO sightings were usually made just before dawn or a few minutes after sunset. They were reported to be shaped like "disks" or "spheres" and to be bright reddish-

FIGURE 3: General Dynamics B-58 Hustler delta-wing jet bomber.

orange, pink, or reddish-white in color. The coloration is caused by the sun's slant rays reflecting off the balloons' surface.

A Lockheed EC-121 Constellation early warning picket aircraft is an example of an unfamiliar type of aircraft which, to the uninitiated observer, might appear to be a UFO.

Many new types of delta-wing aircraft are in operational use. The unusual configuration of this type of aircraft could cause possible confusion and be reported as a UFO.

Lights used on the Boeing KC-135 Stratotanker aircraft for night refueling missions are often mistaken for UFOs in trail formation. The light from a jet tail pipe, under dusk conditions, is also often misidentified.

Other manmade objects and natural phenomenon which have been reported as UFOs include:

* Conventional aircraft observed from unusual angles;
* Modern jet aircraft flying at great speeds and high altitudes;
* Reflections of sunlight, moonlight, and starlight from aircraft and balloons at great heights;
* Searchlight reflections on clouds;
* Meteors, comets and stars;
* Planets observed at certain times of the year;
* Meteorological phenomena;
* Cloud formations;
* Birds, especially migratory formations;
* Dust and haze;
* Kites, fireworks and flares;
* Rockets; and
* Contrails.

What follows is a discussion of some of the factors which may affect the identity of flying objects.

SHAPE

Shape is an important factor in determining the identity of a flying object. Many of the strange shapes of UFOs reported in the past could not be identified in terms of familiar objects, but in many instances could have been reflections from conventional objects viewed under unusual conditions. Light and shadow produce fantastic distortions, especially when viewed objects are at great distances and bathed in varying degrees of darkness.

The four most common shapes of UFOs reported are:
(1) Elliptical or disk shape
(2) Aircraft shape
(3) Cigar shape
(4) Propeller shape

These varieties of shapes are individual reactions to what may have been familiar or conventional objects seen under unusual conditions, or what was created in the mind of the observer by his physiological limitations and psychological responses. Fatigue, unusual weather conditions and stress can induce such manifestations.

One UFO report stated that the object was shaped like a conventional aircraft, but was luminous and surrounded by a red glow. This phenomenon could have been an actual aircraft glowing from an unusual play of moonlight or starlight on its metal parts.

A disk-like object, with illuminated portholes, could be a conventional aircraft distorted in shape and stripped of wings by a temperature inversion mirage effect and reflecting light from windows.

Transparent, cigar-shaped objects, illuminated from the inside and emitting an exhaust, could be jet aircraft at high altitudes where they appear wingless. The mirage effect of a temperature inversion could cause the apparent illumination and transparency.

Saucer-shaped objects, which hover and maneuver erratically, could be the planet Venus or Mars seen near the horizon at certain times of the year. When objects are viewed through haze or mist, or hand-held binoculars, the limitations of the human eye can produce what appears to be a hovering effect, or erratic movement.

Propeller-shaped objects could be conventional or glider-type aircraft, distorted in shape by mirage effects caused by a temperature inversion.

METEOROLOGICAL AND ASTRONOMICAL FACTORS

Scientists have been exploring the mysteries of the universe for many centuries. Today they know a great deal more about the composition of the galaxy which includes the earth among its many planets, stars and other celestial bodies. Yet, many questions remain unanswered and the search for more knowledge in the broad field of astronomy continues. The same is true regarding the earth's atmosphere, and although considerably more is known regarding the natural laws which govern the sea of air around the earth, there are many aspects of meteorology that are not yet fully understood.

It is not unusual for the mind to become confused by unusual astronomical and meteorological conditions transmitted to it by the eye. Thus, the sky has been the setting for many strange sights which were not readily understood.

Under certain weather conditions, reflection and refraction processes can transform conventional aircraft, automobile lights, planets, meteors, and other identifiable figures into UFOs of many shapes and colors. Clouds, haze, industrial smoke, water droplets and ice particles in the atmosphere are typical ingredients which make up atmospheric lenses through which many illusions of flying objects are seen. Car lights reflecting on clouds can create luminous disks which dart erratically through the sky at terrific speeds. Other light sources can produce similar illusions with appropriate variations, many of which even have specific colors provided by refraction of the light through water and ice particles in the atmosphere.

Mirages

One of the most common causes for optical illusions of distorted and displaced objects is the mirage. The index of refraction of the atmosphere varies in the vertical, causing radio and light rays to bend. The prime contributor to index of refraction variability is the temperature variation in the vertical. However, for radio waves, water vapor pressure (humidity) changes are also contributory.

For radio or radar, the ray paths are assumed to be straight lines over an earth whose radius is $4/3$ the true earth's radius under standard atmospheric conditions. This means the radio horizon is farther from the observer than it would be if the rays traveled in straight lines over the true earth.

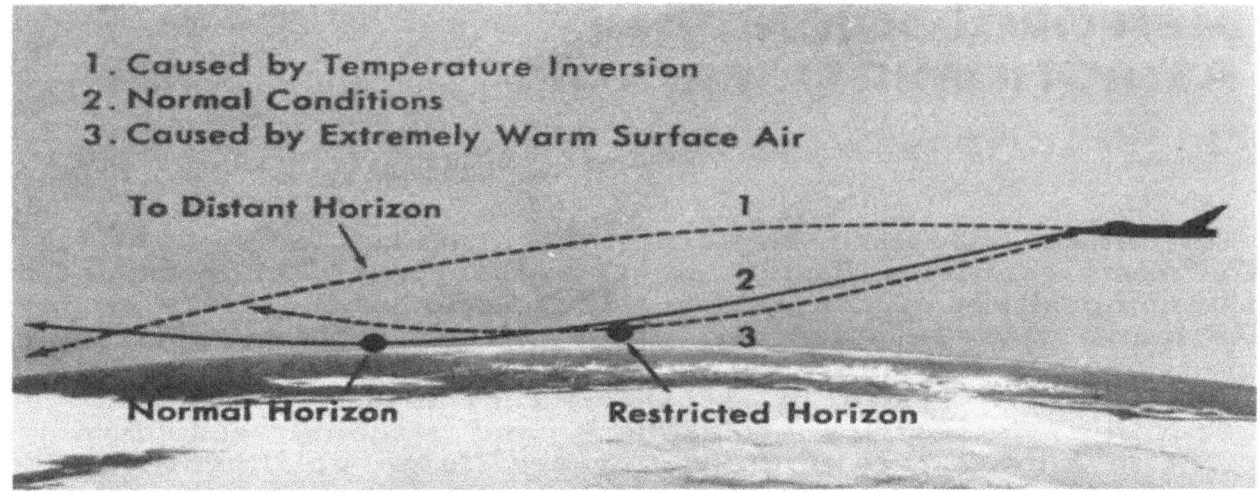

FIGURE 4: Refractive effects.

Light rays follow the same kind of curved path as do radio rays, except that the moisture variations do not contribute to the bending. To an observer or to a radar antenna, the apparent path of the ray is a straight line tangent to the curved path, as in Figure 4. This means an object seen over the true horizon may appear to "float" above the earth. Furthermore, if the object has much vertical structure such that rays travel different paths to the top of the object and to the bottom of the object, the differences in curvature of the top and bottom ray paths may cause a vertical distortion of the true height of the object. One can pursue this subject in atmospheric optics under the phenomena known as "looming."

Refractional Dispersion

Combined refraction and dispersion of the earth's atmosphere can cause a celestial body to appear to be at a different location in space and distort its normal color as well. When the object is low on the horizon, this condition is particularly prevalent. The planet Venus, for instance, may appear as bright red on the bottom and bright blue at the top edge, thereby giving the illusion of a flying object emitting red exhaust trails. An observer flying in an aircraft may easily mistake such an apparition for a flying object. As the aircraft moves through the atmosphere at an advanced speed, its position relative to the object naturally changes and the atmospheric conditions in line of sight between the aircraft's position and the object may change as well. The object thus may assume apparent characteristics of erratic behavior and fantastic shapes and colors.

FIGURE 5: This snow-capped mountain, viewed from an aircraft, is actually a mirage, as described by a pilot who experienced the phenomenon. Under other type mirage conditions, land masses may appear as cigar-shaped flying objects.

Planets

Venus appears brightest of all the planets, with Jupiter and Mars next. Venus, at its brightest, can be seen in daylight. These planets are sometimes "morning stars" and "evening stars." An almanac should be consulted to determine which planet is in the morning or evening sky.

In the past, Venus, Jupiter, and Mars, when low on the horizon, have been observed to change color and move when viewed through haze or mist. If one of these planets is stared at for any length of time without any balancing point of reference, it can appear to perform erratic maneuvers. Thus, the planets of brighter magnitude in our solar system provide a source of illusionary flying objects.

Comets

Comets and meteors have their effect in the field of mistakenly-identified flying objects, although sightings of comets are rare simply because their incidence is so low.

Comets are nebulous bodies revolving around the sun for the most part in long ellipses. Although their periods are very uncertain, a few, such as Halley's Comet, return periodically. The nucleus of a comet, a minute disk of light, strengthens in brilliance the nearer its orbit brings it to the earth. Some comets become bright enough to be seen even in daylight. The long tail of the typical comet is composed of matter repelled away from the sun.

Meteors

Meteors are particles continually entering the earth's atmosphere where they become so intensely heated they turn into incandescent gas.

Meteors will appear to vary in shape, from round to elongated teardrops, and in size from tiny pinpoints to the size of the moon. Colors will range from yellowish-white through red, to blue and green hues depending on the atmosphere.

Although often observed singly, meteors may be observed in clusters. The time in sight is generally less than 10 seconds. Meteoric bodies themselves cannot be picked up on radar; however, the meteor trails are generally good reflectors and will often "paint" on radarscopes. One or two sporadic meteors per hour can be observed at any single locale.

Bright meteors are known as fireballs. The ones which penetrate the lower parts of the atmosphere, where they explode with a noise like distant thunder, are called bolides. These are rare—probably no more than a few dozen appear during an average year. When a meteor, of such size that it is not entirely consumed by frictional heat after it enters the atmosphere, collides with the earth's surface, it is called a meteorite. It is estimated that about 2,000 of these enter the earth's atmosphere during an average year.

The appearance and behavior of meteors streaking through the earth's atmosphere take on various fantastic forms, depending upon their size and composition and the meteorological conditions through which they are viewed. A meteor with the brilliance of the North Star (Polaris) can be caused by a particle no larger than a grain of sand. A particle no bigger than a pea can become a fireball. Examination of meteorites reveals that most are irregular in shape; however, many become conically shaped in their passage through the earth's dense atmosphere.

Meteors may appear as bright balls or disks with fiery tails, which could be mistaken for jet or rocket-type exhausts. It is not uncommon for meteors to appear as flaming fireballs, with colors ranging from dull red to bright green, and they may even travel in clusters, giving the appearance of flying objects in formation. Meteors also may move relatively slow and appear to follow a path parallel to the horizon, thereby giving strength to the illusion of flying objects.

Large meteors have long paths and may cross from one horizon to the other in the view of one observer and pass far beyond. Those meteors overtaking the earth during evening hours may travel initially as slowly as seven miles per second, while those meeting the earth's rotation head-on during morning hours can be traveling more than 40 miles per second.

Each year, the earth passes through certain meteoric showers at specific times, which invariably results in a large number of UFO reports. These annual showers can be exceptionally brilliant, having thousands of meteors and meteor trails. Some of the most prominent of these are the Perseids (August), Orionids (October) and the Leonids (November). The Leonids, for example, which last for approximately seven days and reach a maximum about November 16, have provided close to 200,000 meteors between midnight and dawn. Any good astronomical text will furnish the dates and schedules of these periodical showers.

Sundogs and Moondogs

The reflection of the sun in a layer of flat ice crystals can cause a phenomenon known as a sub-sun, commonly called a sundog. This apparition will appear at a point adjacent to the real sun and can be as brilliant as the sun itself. The sub-sun can develop a pattern of other sub-suns, causing a further complicated illusion. At night, the moon will reflect in the same manner under like meteorological conditions. This type of apparition is particularly discernible from aircraft at high altitudes.

The size and brilliance of sundogs and moondogs, and their behavior in relation to the observer, will depend upon the loca-

FIGURE 6: Shining object was photographed by individual flying over Ind., in 1954. It was identified as a "sun dog," or refraction of the sun's rays, a phenomenon similar to the one producing a rainbow.

tion and density of the reflecting source, e.g., ice or frost-crystal formations, and, of course, upon the position and movement of the observer.

Cirrus cloud formations are effective viewing screens for illusions resulting from reflected or refracted light, as they contain ice crystals. These clouds exist in the upper atmosphere, so that conditions are favorable throughout the year for sundog and moondog apparitions. However, such phenomena usually are discernible at lower levels only during winter months in temperate zones.

Auroras

The aurora borealis, or northern lights, produces conditions and phenomena which have been associated with mistakenly-conceived flying objects. Auroral activity is associated with the earth's magnetic fields and other solar activity.

The maximum auroral zone in the northern hemisphere follows roughly a circle around, and about 23 degrees of latitude away from the magnetic pole. Auroras are seen only infrequently below 45 degrees latitude.

The aurora borealis cannot be seen in daylight, and during moonlit periods it is inconspicuous. It is sometimes bright enough to cast easily discernible shadows; its surface brightness surpasses even that of the moon. The most distinctive form of the aurora is that of a curtain or long wavy band, often with folds and flutings in it. Although the lower edge of the aurora is nearly horizontal, the band would appear as an arc, due to its great distance from the observer. Auroras may consist of more than one curtain and may appear and disappear rapidly, remain constant for long periods, or move slowly across the sky. Some may appear merely as formless, diffused lighting in the sky. Faint auroras may appear colorless. Bright auroras are usually yellow-green, but other colors such as red, blue, grey and violet sometimes appear. A yellow-green curtain often will be tinged with red around its lower edge. Auroras may appear high in the sky or low on the horizon, depending on the distance of the particular phenomenon from the observer.

While the chances of the aurora borealis itself being mistaken for a flying object are remote, the erratic lighting conditions it produces may often be a contributing factor to a sighting.

Artificial auroras are created by scientific rockets carrying sodium or barium to seed the upper atmosphere. This upper atmosphere light starts out as spherical in shape and, after several minutes, becomes elongated and shredded by winds and turbulence. The colors appear as white, green, red and sometimes yellow. Scientific rockets seeding the upper atmosphere have been launched from Churchill Research Range, Manitoba, Canada; Wallops Island, Va.; and Eglin Gulf Test Range, Fla. Under clear conditions, the light from the seeding chemicals can be seen up to 600 miles away.

RADAR SIGHTINGS

In certain instances, unidentified objects have been reflected on radarscopes, both ground and airborne. Generally speaking these radar sightings fall into explainable patterns and are caused by certain meteorological phenomena, or familiar objects that are observed under unnatural circumstances. Spurious returns may also be caused by mutual interference between radar sets, jamming, and equipment malfunction.

Radar echoes can be produced by a variety of objects, many of which are not visible to the human eye. A majority of solid objects which return radar energy produce responses on the radarscope that are easily recognizable. Moving objects, such as birds, aircraft and meteorological balloons, are normally recognizable by their size and velocity. However, some balloons, such as ionospheric balloons, ascend to altitudes above those of normal aircraft and travel with the upper air currents, sometimes at speeds above 100 m.p.h. Radar returns from these balloons could be misinterpreted as UFOs.

Certain meteorological and astronomical conditions will present radar returns that are unusual. Radar waves must travel through the earth's atmosphere where, like light waves, they may be bent by unusual temperature and moisture conditions. Radar waves may be refracted or reflected by atmospheric conditions to where ground objects may seem to represent an aircraft or flying object. Even with a moving target indicator, reflected images of distant ground objects may appear to be moving because of the movement of air layers.

Temperature inversions, in which a cold air mass is overlaid by a warmer air mass, can greatly increase the distance from which normal radar returns are received. Thus objects may appear to be much closer than they actually are and these distant objects, superimposed on the normal radarscope picture, may result in misinterpretation and confusion.

Radar echoes may be produced by condensed water vapor in the form of raindrops, ice crystals, or snow. These radar reflections may cover a wide area which has diffused, irregular

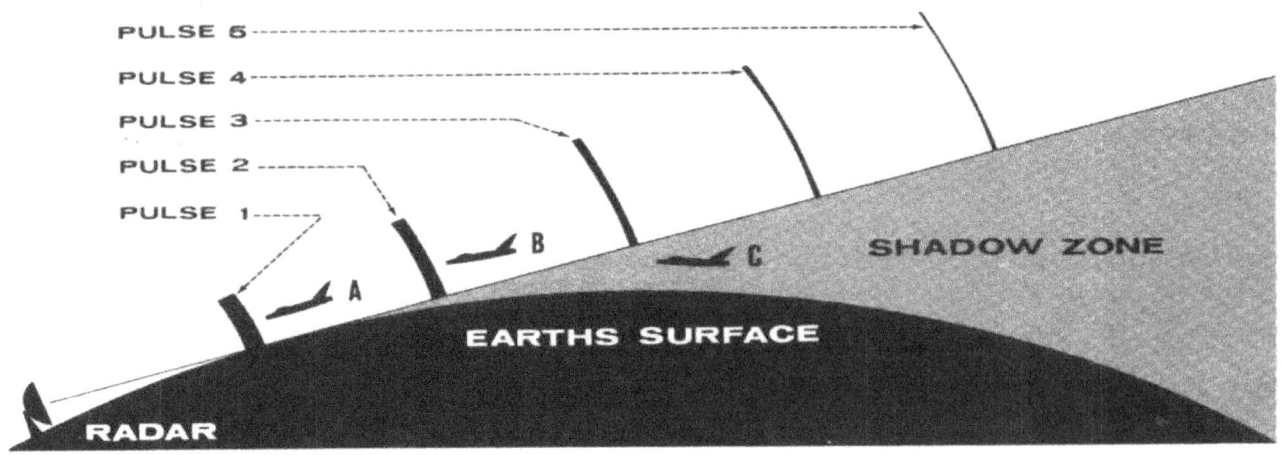

FIGURE 7A: The transmission of a radar pulse, under normal atmospheric conditions, follows line of sight. Therefore the curvature of the earth would place Target "C" in the shadow zone.

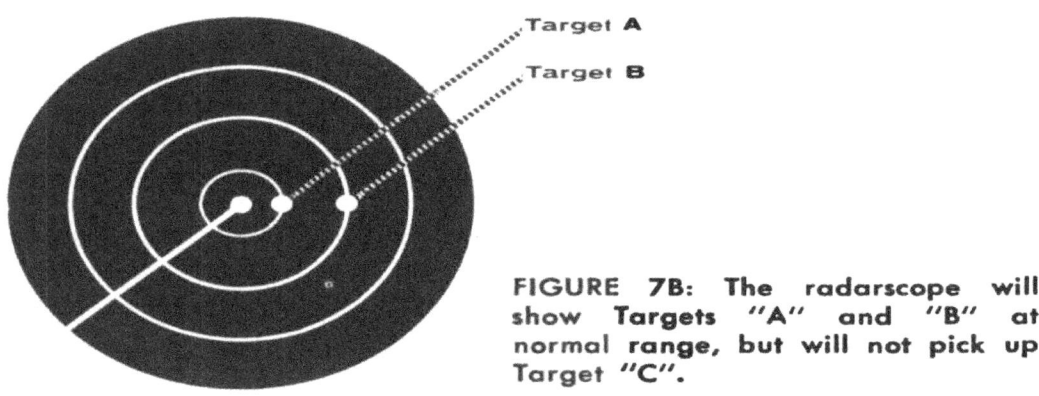

FIGURE 7B: The radarscope will show Targets "A" and "B" at normal range, but will not pick up Target "C".

boundaries and fluctuating intensities. Movement of this water vapor will be determined by the movement of upper air currents. Normally, these patterns are easily recognizable by their size and radar return; however, they may appear confusing and result in false interpretations.

Meteors that enter the earth's atmosphere and get within range of radar may cause reflections that are extremely difficult to verify. Radar responses to these meteors may occur at any range or altitude, depending only upon the capabilities of the radar set. Radar reports resulting from this type of phenomenon can be verified by a study of the expected paths of meteors at the time of the incident. It is the ionized trail of the meteor which is most often seen on radar. The trail moves with the wind.

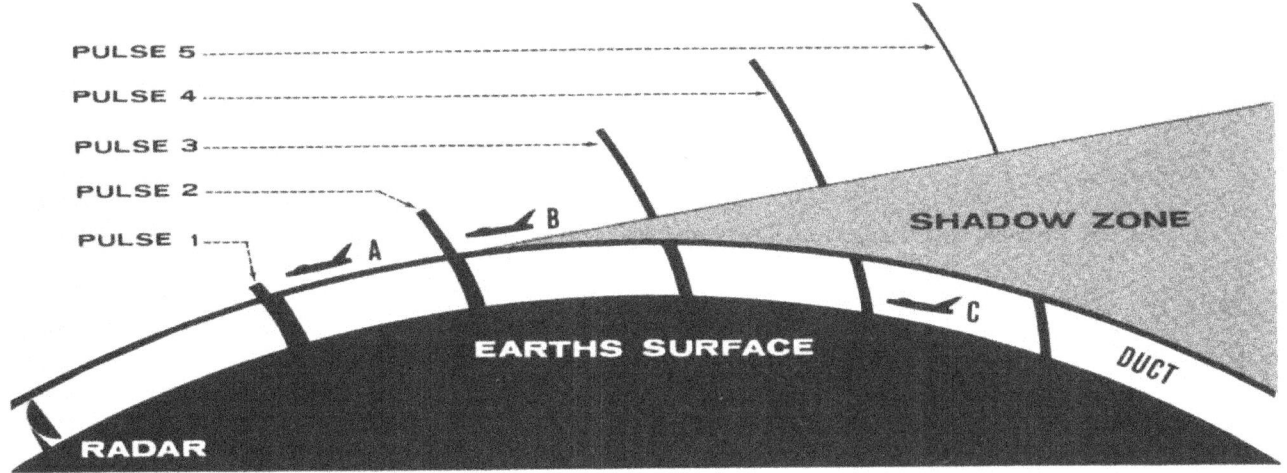

FIGURE 8A: Under abnormal conditions, with cool air overlaid by a warmer air mass, a duct is formed through which the radar pulse travels and reflects Target "C" at a much greater distance.

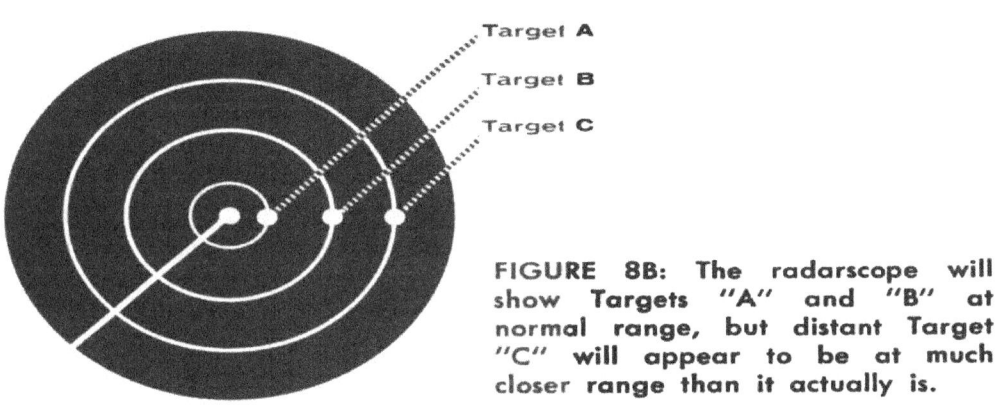

FIGURE 8B: The radarscope will show Targets "A" and "B" at normal range, but distant Target "C" will appear to be at much closer range than it actually is.

In addition, there is the possibility that one radar set, which has characteristics similar to those of another radar set within range, may cause interference and unusual responses that could lead to confusion and inaccurate interpretation. Although this type of interference may cause the appearance of one or even two targets on the radar screen, it can generally be recognized quite easily.

Experience in the operation of radar will provide the operator with the ability to recognize most unusual phenomena when they occur. However, occasionally verification of meteorological or astronomical data may be necessary to identify what otherwise might be considered a UFO. When an operator accustomed to stable atmospheric conditions is tranferred to an area where the atmosphere is unstable and turbulent, he must be made aware of possible anomalous propagation problems.

PHYSIOLOGICAL ASPECTS

Physiological factors may have profound effects upon an individual's ability to observe and to interpret observations accurately. One of the greatest hindrances to human understanding can result from deception of the senses. The sense of sight is, by itself, purely a physical process, and the perception and understanding attached to visual sightings is determined largely by memory of past experiences and familiarity with surrounding objects. This relation of experience to the interpretation of visual sightings permits many errors.

Perceptual error may be applicable particularly to aircrew members operating high-performance aircraft, under adverse or unusual weather conditions, under tension, and during periods of extreme fatigue.

The aircrew member is generally familiar with many of the unusual observations associated with meteorological and astronomical phenomena. However, many unusual observations are the result of certain physiological effects that may be unknown or unfamiliar.

Occasionally, objects that exist on the surface of the eye may be mistaken for distant objects. These objects take various forms. Tiny specks of dirt may appear as shimmering globules of light and, if a speck is illuminated by an outside light source, it may appear as a large, out-of-focus blob of light. If this speck is viewed against a dark sky or background, it may be quite spectacular. As this speck floats across the pupil of the eye, it will create the appearance of movement.

Many reported UFOs, described as flying saucers, glowing disks, shiny spots or a string of pearls, are nothing more than minute blood capillaries on the surface of the retina of the eye, or tiny corpuscles, which become visible under special conditions of illumination.

Another physiological phenomenon is that of after-image. Rhodopsin, or visual purple, is associated with dark adaptation of the retina, and sudden light destroying the visual purple affects the retina of the eye and causes a dark image to remain visible for some time after the light has been extinguished. Flashes of lightning, comets, or meteors will cause this effect and may be confused and interpreted as UFOs.

Hypoxia, resulting from lack of oxygen, has varying effects on the ability to react and to observe accurately. The effects of hypoxia may vary much in the same manner as those of alcoholic intoxication. Usually vision is affected, reactions are retarded and observations are distorted. An oxygen mask leak may cause alternating stages of hypoxia and normalcy, with the individual often being unaware of these changes.

In a series of tests conducted at the USAF School of Aerospace Medicine to determine the effects of fatigue, it was discovered that extreme fatigue may cause an individual to hallucinate, imagining that he sees a variety of unusual objects with a vividness to make them seem quite real. Fatigue, even in minor degrees, will slow down reaction time and reduce ability to observe and interpret observations.

Two phenomena that occur frequently are those of autohypnosis and autokinesis. In both of these reactions, a stationary light will assume apparent movement. In autohypnosis, this reaction is caused by continued attention to an external light source. Autokinesis is the result of observing a stationary light under circumstances in which relation to familiar objects is absent.

There is strong evidence that a great many visual problems, both physical and physiological, arise as a direct result of flight at high altitudes.

When flights are conducted at relatively low altitudes, the visibility of distant targets will be reduced by atmospheric haze. This is because light emanating from objects in space is gradually attenuated by absorption and by primary and secondary scattering along the pathway of sight. Along with the variation of the contrast by atmospheric interference, there is a shift of the apparent contours. This has been disclosed by experiments performed at the USAF School of Aerospace Medicine. From these studies, it was concluded that the apparent angular size and apparent distance of objects depend on the brightness reduction by the atmosphere. With increasing altitudes, the deviation of the apparent luminance from the actual luminance of an object in space will result in the object appearing brighter than it actually is. This may result in false identification of a normally familiar object.

PSYCHOLOGICAL FACTORS

Reasoning ability, degree of susceptibility to suggestion, and general mental attitude are vital factors in identifying and reporting flying objects. Failure to note details accurately and a tendency to overdraw descriptions of sightings can result in failure to identify. An over-active imagination, coupled with physiological strain, can transform unfamiliar meteorological or astronomical phenomena and light aberrations into UFOs.

Perception and feeling are closely related and can have a marked effect upon understanding. Motivation in many instances determines how we interpret what we see, and expectancy can induce manifestations which are only indirectly related to actual physical phenomena or objects. The separation of what may be observed through the senses from what is known through thought or intuition is difficult, inasmuch as understanding is derived from a combination of both. However, an objective attitude, which permits assessment of observed characteristics, rather than suppositions or theories, will assist the observer in avoiding distorted descriptions.

It has been suggested that the world each of us knows is a world created in large measure from our experience in dealing with our environment. When two points of light, one brighter than the other, are placed at an equal distance from an observer in a dark room, the bright point of light looks nearer than the dim light, if one eye is closed and the observer remains motionless. The direction from the observer, as well as difference in brightness, will result in an apparent variance in distance. Should two equally bright lights be placed near the floor, one about a foot above the other, the upper light will appear to be at a greater distance from the observer than the lower one. Conversely, when the lights are placed near the ceiling of the room, the lower light will appear to be farther away.

When two partly inflated balloons are illuminated indirectly and fastened in positions about one foot apart, where their relative brightness and inflation can be controlled, the observer will experience a variety of reactions as to what he saw.

If the brightness and size of the two balloons remain the same and the observer views them with one eye at a distance of approximately 10 feet, he sees two bright spheres equidistant from his position. If the relative sizes are changed and the brightness remains the same, the larger balloon usually appears nearer. When the size is changed continuously, the lighted balloons seem to move back and forth, giving the effect of erratic movement of lighted spheres through space. This is true even when observed with both eyes. If there is a variation in relative size and brightness, most observers are inclined to judge distance by relative size rather than by relative brightness.

The effect of these tests upon the observer is premised on the fact that he draws upon past experience in assessing distance based upon relative size and brightness. He assumes that, since the two points of light appear similar, they are identical and of equal brightness. Therefore, the point of light which seems brighter must be nearer. In the case of the two points of light placed one above the other, past experience leads the observer to assume that, when he looks down, the lower light is nearer and, conversely, that, when he looks up, the higher light is nearer.

With regard to the seeming variance in distance when the size of objects is changed continuously, rarely has the observer seen two fixed objects at the same distance change in size. Usually any change in size of an object results from a change in the position of the object in relation to the position of the observer. As the object draws nearer, it becomes larger, and the reverse is true as it draws farther away. Therefore, in the case of the two balloons, the observer assumes that any change in size of the two balloons results from a variation in distance from his point of observation.

These experiments show how misinterpretations can result from the relation of visual perception to past experience in an effort to understand and recognize the object or objects seen.

When we see an object, we derive an impression not only of its location, but also of its existence as an object, and the location as related to visual perception will color the characteristics it possesses. Objects seen through haze or mist, or in reflected light, will assume characteristics they do not possess normally, but, because they have been perceived visually, the observer tends to accept them as real. Thus, psychologically, he creates an object with characteristics which do not exist. It is essential, therefore, that the observer analyze his observations in relation to unusual weather or lighting conditions and reject characteristics which deviate from the normal and can be explained by the unnatural conditions under which they were seen.

When we see an unfamiliar object, we draw upon our individual past experience in an attempt to identify it. If the unfamiliar characteristics of the object cannot be related to past experience, we have a feeling of uncertainty and it is then that we draw upon imagination in an effort to relate visual perception to understanding. Imagination is colored by suggestion and herein lies an inherent danger.

We are open to suggestion constantly in our daily lives. Advertising media, artists' conceptions, modern-day science fiction, propaganda, exaggerated film versions, publicity on perpetrated hoaxes, and the imaginings of zealots and fanatics all react upon the consciousness in the form of suggestion. When we seek an explanation for the unusual or unfamiliar, and attempt to draw upon imagination instead of rational analysis, suggestion influences our thinking.

Physiological changes due to fatigue and intense strain enhance the susceptibility to suggestion and may induce psychological manifestations which a more rational state of mind would reject. The observer should attempt to evaluate his observations. Objective analysis of those characteristics he has observed, in relation to the conditions under which they were seen, will assist in identification of the unfamiliar object and result in more accurate reporting.

VISUAL PERCEPTION

Since visual perception supplies the first awareness of a flying object, it is important to know "how to see." Knowing "how to see" will facilitate identification and reporting of flying objects. The following aids to "seeing" to the best advantage are provided from Air Force Pamphlet 160-10-3, "Your Body in Flight."

Sky Search

It is a common misconception that the eye "takes a picture" of everything within its field of view. This is not true. Pick out any word in this sentence and then move your eye to the next and then the next. You will discover that you can no longer read the first word after having moved your eye about 5 degrees.

You see best in daylight and the eye sees by moving in short jumps. It is not a sweeping but a jerking motion with which you see details around you. This is of the utmost importance to the combat pilot scanning the sky for the enemy. Experiments have shown that the eye sees nothing in detail while it is moving. It sees only when it pauses and fixes an object on its retina. In scanning the sky, do not deceive yourself that you have covered an area with a wide, sweeping glance. The correct way to scan is to cover an area with short, regularly-spaced movements of the eye.

Depth Perception

Judgement of distance is done subconsciously in a combination of ways: Close up, we depend on binocular vision, each eye seeing an object from a different angle. At distances beyond binocular range, which is usually the case in flight, we judge it on a one-eye basis. Examples of methods of depth perception are given in **Figures 9A through 9G**.

Accommodation

The eyes change focus to see objects within about 20 feet, but do so very little for distant objects.

Seeing at Night

It is easy for your eyes to play tricks on you at night when you stare for some time at a light—say, the tail-light of a lead airplane. What happens is technically known as autokinetic movement, or more commonly as stare vision. If the light is

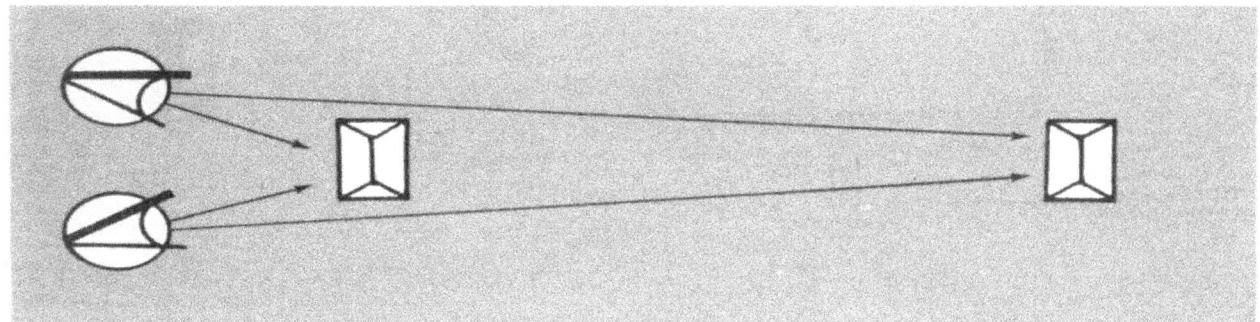

FIGURE 9A: Binocular vision.

FIGURE 9B: From the known size of an object and how much of our visual field it fills.

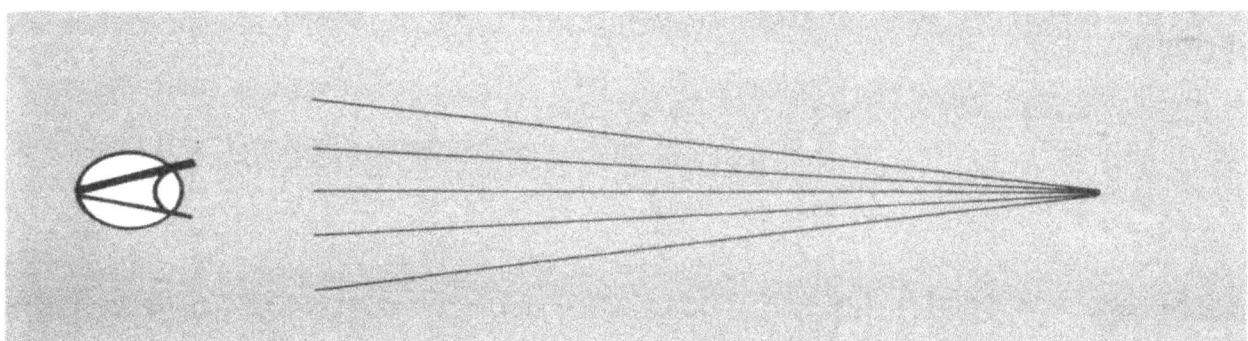

FIGURE 9C: From our knowledge of perspective and the convergence of parallel lines at a great distance.

FIGURE 9D: From overlapping—an object overlapped by another is known to be farther away.

FIGURE 9E: From light and shadow—an object casts a shadow away from the observer if the light is nearer.

FIGURE 9F: From aerial perspective—large objects seen indistinctly apparently have haze, fog, or smoke between them and the observer and, therefore, are usually at a great distance.

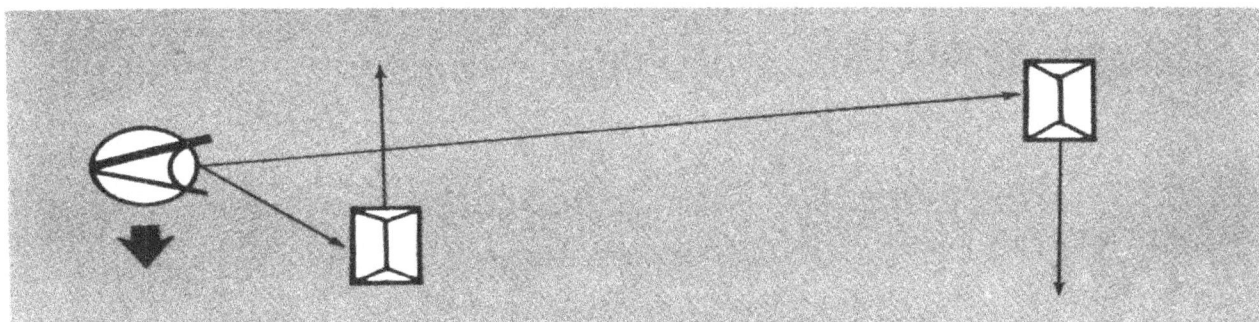

FIGURE 9G: From motion parallax—When the observer fixes his sight on one object while his head or body moves, other objects apparently moving in the same direction as he are judged to be more distant, while those apparently moving in the opposite direction are judged to be nearer.

stationary, it may seem to move and swing in wide arcs. If the light is moving, it may seem to move to the side when it is actually going straight ahead. The cure for stare vision is don't stare—keep shifting your gaze from point to point.

Another common illusion at night is to see a light expanding or contracting at a fixed distance from you when actually the light is approaching or going away. Again, shift your gaze.

One last tip on seeing at night is to keep your windshield scrupulously clean. Dust, grease, water droplets, scratches, and the like all obstruct your view, night or day. Many a speck on the windshield could, after a few hours, take on the silhouette of a UFO.

With regard to color perception at night, blue and green lights are seen most easily; red and orange are seen least easily. However, at night, red is most readily identified as being red. Other colors, while able to be seen at greater distances, tend to become washed out, or white in appearance.

QUESTIONS AND ANSWERS

1. What is a UFO?

A UFO (unidentified flying object) is an aerial phenomenon or object which is unknown or appears out of the ordinary to the observer.

2. Are there really any "flying saucers?"

The term "flying saucer" is really a science fiction term coined in recent years. No evidence has been submitted to or discovered by the Air Force that *identified sightings* represent technological developments or principles beyond the range of our present scientific knowledge. There has been no tangible evidence to indicate that unidentified sightings are extraterrestrial vehicles under intelligent control. No UFO has been determined to represent a threat to our national security.

3. Do UFOs have any particular size and shape?

UFOs have been reported to have a variety of shapes and sizes. There appears to be little pattern as to shape, size or color.

4. How fast are UFOs supposed to go?

There have been reports which estimated the speed of these phenomena at 25,000 m.p.h. No evidence exists that intelligently directed machines other than our own (earth) rockets travel at such high speeds.

5. Is there any sound emitted from UFOs?

Some observers have reported hearing sounds; others have reported no sounds associated with the sightings. The sounds reported have varied as much as have reported shapes and colors.

6. Have UFOs been photographed? If so, have negatives been sent to the Air Force?

Photographs of phenomena reported as UFOs have been submitted to and evaluated by the Air Force, which welcomes bona fide photographs. However, original negatives are required for accurate analysis. (Negatives are returned to the owners after evaluation.) Of those photographs studied (made from original negatives loaned to the Air Force), none has proved the UFO to be of extraterrestrial origin.

7. Is there any time of the year when UFOs are seen the most?

Between early spring and the beginning of winter. This is attributed to the fact that more people are out-of-doors during the warmer months.

8. Has any pattern been established by reported UFO observations?

No definite pattern of shapes, sizes, colors, speeds, etc., have been established. About the only pattern noted, other than seasonal changes in the number of sightings, is that UFO observations appear to be directly related to weather conditions. When the weather conditions are such that more people are out-of-doors, there are more sightings.

9. How many UFOs are sighted by day? By night?

No count has been made, but the vast majority are night sightings and consist of lights in the sky, including stars, planets, meteors and aircraft lights.

10. Why do stars flash and flicker?

The earth's atmosphere, especially in the lower dense layers, has a turbulent structure. This causes light rays from stars to be refracted and to appear to twinkle. This is why astronomers prefer to locate observatories on mountain tops, above the most dense layers, or in areas where the characteristic turbulence is minimal.

11. Why do artificial satellites seem to move in tiny zigzags when crossing the sky?

To the naked eye a bright artificial satellite frequently appears to move across the sky in a zigzag course rather than in a smooth arc. The familiar illusion has been ascribed to the fact that our eyes do not move continuously but in short jerks.

12. What are the comparative distances involved in traveling from outer space?

The nearest star like our Sun is Proxima Centauri, which is over four light-years away or more than 25 trillion miles. Light travels at 186,000 miles per second or nearly 670 million miles per hour. At this speed you could make about 8 trips around the Earth in one second. A trip to the Moon would take one and one-half seconds and a trip to the Sun would require 8 minutes. Traveling to Pluto, our solar system's outermost planet, would take 5½ hours, and a flight to the nearest star, Proxima Centauri, would require more than 4 *years*.

13. Are UFO sightings new?

Recorded sightings of UFOs are not new, although descriptive terms such as "flying saucers" and "UFOs" have been coined in recent years. As far back in history as 593 B.C., Ezekiel recorded seeing a fiery sphere. What we now know as "Northern Lights" has been the cause of numerous UFO sightings. St. Elmo's Fire, well known to aviators and sailors, also has been reported as UFOs. St. Elmo's Fire is merely atmospheric electricity which creates strange effects when viewed under unusual conditions.

During World War II pilots reported balls of light, orange or red, but without apparent structure, which followed their aircraft. Seen during daylight and dark and in good and poor weather, these phenomena, dubbed "Foo Fighters" by the pilots, may have been reflections of the aircraft themselves somehow distorted by unusual atmospheric conditions or some electro-magnetic or electrostatic phenomenon similar to St. Elmo's Fire.

14. Are UFOs reported by unreliable, unstable and uneducated people?

Yes. But UFOs are reported in even greater number by reliable, stable, and educated people. The most articulate reports come from obviously intelligent observers.

15. Are UFO "buffs" the only ones who report UFOs?

No. The exact opposite is much nearer the truth. Only a negligible handful of reports submitted to the Air Force are from "true believers."

16. Are UFOs ever reported by scientifically trained people?

Yes. Some of the best, most coherent reports have come from scientifically trained people. They usually request anonymity, which is always granted. The quality of the source does not mean disagreement with Air Force findings—usually these people readily accept a finding when additional facts are made known to them.

17. Are UFO reports generated by publicity?

While there is normally an increase in the number of UFO reports when sightings are widely publicized, publicity cannot be credited as the sole cause of increased reports over a given period.

18. Who is responsible for investigation of UFO reports and how successful are these investigations?

The U. S. Air Force is responsible for investigating UFO reports. Since the inception of the Nation's program of UFO investigation, only 6.4 percent of the reported cases have remained unidentified. This statistic is not completely representative of the Air Force's success with the program, for a great majority of the unidentified cases occurred during the first 5 years of the project. From its inception in 1947 until 1952, 25.3 percent of the reported sightings were unidentified; from 1953 to 1957, 4.4 percent were unidentified; from 1958 to 1962, 2.5 percent were unidentified.

19. What is Project Blue Book?

"Blue Book" is the nickname of the Air Force project established to conduct a program of investigating reports of unidentified flying objects (UFOs), of evaluating those reports, and of responding to observers on the evaluations reached. It is normally limited to the territory within the jurisdiction of the United States. Observers may report sightings to the nearest Air Force base or directly to: Project Blue Book, Wright-Patterson Air Force Base, Ohio 45433. Inquiries concerning "Blue Book" should be directed to: Project Blue Book, Headquarters, U. S. Air Force (SAF–OICC), Washington, D. C. 20330. News media representatives should contact: Office of Information (SAF–OIPC), Office of the Secretary of the Air Force, Washington, D. C. 20330.

20. What general emphasis does the Air Force place on investigating UFO reports?

The Air Force is primarily responsible for providing aerospace defense. To determine whether there is a threat to national security and to discover the stimuli which cause sightings, the Air Force uses the services of its trained scientists and technicians to evaluate observations of unusual aerial objects over the United States. In addition, the University of Colorado has been selected by the Air Force to conduct independent investigations into UFO reports, to analyze phenomena associated with UFO sightings and to make recommendations on the Air Force's methods of investigating and evaluating UFO reports. The Air Force's UFO files, as well as any other information in the possession of the Air Force, are made available to the university and its investigation team. Additionally, all Air Force installations within the United States assist the team if requested. The investigators, however, conduct their research independently of and without direction from the Air Force.

21. What is swamp gas?

Swamp gas is formed from decaying vegetation and consists of methane (CH_4) and phosphine (PH_3). Phosphine can be ignited spontaneously when exposed to air. Methane in air is dangerous when a flame is present. Phosphine does not burn with a hot flame, but is luminescent.

Several terms have been given to swamp gas over the centuries. The Romans called it *Ignis Fatuus*, or fool's fire, because travelers at night were lured off their paths by it into swamps and marshes, thinking it came from a dwelling. Other terms for it are *Friar's Lanthorn* and *Will-o'-the-Wisp*. Russian superstitition holds that swamp gases are actually spirits of still-born children which flit between heaven and hell. Shakespeare made references to it in *Henry IV, Part I (Act III, Sc 3)*.

22. Does the Air Force have any evidence that pilots and airplanes have disappeared while checking out reports of UFOs?

No. There has been one case, however, in which a pilot lost his life while investigating a UFO which was later identified as a U. S. Navy "Skyhook" balloon. This occurred at Godman Field, Ky., in 1948.

23. Has anyone ever been hurt as a result of confronting a UFO?

There have been cases when persons reported being burned or felt intense heat from UFOs. However, in no instance has there been any tangible evidence to substantiate the reports.

24. Have our astronauts reported anything that could not be explained?

The National Aeronautics and Space Administration has reported that there were occasions during Gemini flights when objects were sighted by U. S. astronauts, but in all instances, these objects were later identified as satellites or parts of satellites launched into orbit from earth.

25. Does the Air Force say that extraterrestrial life does not exist?

No. The Air Force takes no stand on whether or not extraterrestrial life could or does exist. Scientists believe that life could exist on planets other than our own.

26. Is any other Government agency engaged in investigation of UFO reports?

No. Since 1947 this responsibility has rested primarily with the Air Force. Although other agencies have exhibited varying interest in the Air Force project from time to time, no other Government agency now has, or ever has had, direct responsibility for the program.

27. What has been Dr. Hynek's role in the Air Force's UFO project?

Dr. J. Allen Hynek is director of the Dearborn Observatory and Chairman of the Department of Astronomy at Northwestern University. He has been associated with the Air Force UFO program since its inception in 1948 when Project Sign was instituted. He assisted with Project Grudge and then with Project Blue Book Report in 1955, and subsequently became the official consultant to the Air Force on UFOs, the position he still holds today. Probably no one outside the Government knows the subject better than he.

28. Why did the Air Force select the University of Colorado and Dr. Edward U. Condon to conduct the UFO study?

Informal approaches were made to organizations and individuals in the scientific community to obtain advice on leading institutions and experts in this field. The University of Colorado and Doctor Condon were highly recommended by several individuals and agencies within the scientific community.

The University of Colorado, Boulder, Colo., was selected because of its reputation for excellence in the sciences. Boulder, Colo., possesses a unique combination of research institutions specializing in studies of the atmosphere. Among these are the Joint Institute for Laboratory Astrophysics, National Center for Atmospheric Research, and research laboratories of the Environmental Science Services Administration.

Doctor Condon is a physicist with an outstanding background and accomplishments in broad areas of research and research management. He has served as president of the American Physical Society, the American Association for the Advancement of Science, and the American Association of Physics Teachers. He was scientific advisor to the Special Senate Committee on Atomic Energy (79th Congress), associate director of research for Westinghouse Corp. and Director of the National Bureau of Standards. In addition to his teaching and working with scientific societies, Doctor Condon is the editor of one of the leading physics journals, "Review of Modern Physics."

29. Why is the University of Colorado planning to include psychologists in its study of UFOs?

The composition of the study teams by the University of Colorado is a discretionary matter with the University. However, there are valid reasons to include such specialists. Experts with experience and training in studying perception and interpretation, pattern recognition, optical illusion, and optical distortion can study the interaction of the physiological mechanisms of human beings and their effect on the interpretation of physical stimuli. For example, a person who is nearsighted and has astigmatism might give a different interpretation to the same visual object seen by another person who is farsighted and color blind, or one who has normal vision.

Psychologists, as experts in information processing, can translate the information supplied by observers who often are not scientifically familiar with terms useful and meaningful to the physical scientists in the study.

And, of course, experts in social psychology could study group reactions to unusual stimuli. A possible example is rumor phenomena, such as Orson Wells' radio presentation of H. G. Wells' *War of the Worlds* in the late 1930's.

There is no indication of a need to study the mental health of observers or to give credence to explanations of hallucinations, which have been extremely rare during the two-decade Air Force study.

30. What is the Robertson Report?

In 1953 the Central Intelligence Agency (CIA) convened a panel of eminent civilian scientists, headed by Dr. H. P. Robertson, to study UFO reports and to determine whether these phenomena constituted any threat to the security of the United States. Their statement upon completion of the study was released at the time. It reads as follows:

As result of its considerations, the Panel concludes: That the evidence presented on Unidentified Flying Objects shows no indication that these phenomena constitute a direct physical threat to national security."

On Oct. 11, 1966, the CIA, in response to questions on its activities in this regard and on the classification of the bulk of the Robertson report, had this to say:

It is indeed true that in the early 1950's the Central Intelligence Agency was actively studying the matter of Unidentified Flying Objects because at that time there was no way of knowing whether such objects might be originating from sources overseas.

All materials concerning UFOs were subsequently declassified and made available to scientists, scholars and others interested. Materials which were not declassified had nothing to do with UFOs but with the organization and methods of the CIA.

The matter of UFOs currently is the responsibility of the Air Force and CIA has no interest either in building up or debunking any information or views concerning UFOs.

31. What is Project Blue Book Special Report #14?

This report, dated May 5, 1955, reported conclusions reached after study and analysis of UFO reports dating back to mid-1947. The report was unclassified and released to the public. Copies of it are exhausted except for single file copies in offices of interest. Its 309 pages will be reproduced upon request at a cost of 25 cents per page to the requestor. Special Reports #1–13 were administrative progress reports and do not contain any technical data regarding the study of UFO cases.

32. What were Projects Grudge and Sign?

Projects Grudge and Sign were previous designations of Project Blue Book. They were originally classified, but were downgraded to unclassified after their completion. Study copies of these reports have been exhausted. File copies on hand can be copied upon request at the expense of the requestor. The charge is 25 cents per page. There are approximately 450 pages in the two publications.

33. What is the relationship between the Air Force's *Project Blue Book* and the various private organizations interested in UFOs?

There is no direct relationship between the many private UFO organizations and the Air Force's Project Blue Book.

The Air Force does not maintain a listing of these private organizations.

34. Has the project on UFOs ever had a Top Secret classification?

No. In the first few years the project carried a security classification. It was felt that the UFO observations were of experimental or other advanced aircraft not of U. S. origin. As time passed, this assumption was not substantiated and the project's classification was dropped.

35. Why are some reports of sightings by military personnel classified?

Reports are classified only when they involve national security such as the performance of military equipment (aircraft, radar, mission and location of certain installations). For example, if a UFO report had been made by the pilot of the Lockheed YF–12A before it was revealed to the public, any part of that report which might have revealed the existence of the YF–12A would have been classified, but the rest (pertaining to the UFO) would have been unclassified and released.

Bibliography

BOOKS

AUTHOR	TITLE
Arnold, Kenneth and Ray Palmer	The Coming of the Saucers
Beckley, Timothy G.	Inside the Saucers 1962
Brasington, Rev. Virginia	Flying Saucers in the Bible
Davidson, Leon	Air Force Project Blue Book Special Report No. 14
Drake, Eugene H.	Life on the Planets
Edwards, Frank	Flying Saucers, Serious Business

Fuller, John G.	Incident at Exeter
Girvin, Calvin	The Night Has a Thousand Saucers
Girvin, Waveney	Flying Saucers and Common Sense
Hall, Franklin	The Riddle of the Flying Saucers
Heard, Gerald	Is Another World Watching
James, Trevor	Spacemen—Friends and Foes
Keyhoe, Donald	The Flying Saucer Conspiracy
	Flying Saucers: Top Secret
Lorenzen, Coral	The Great Flying Saucer Hoax
Maney, Charles and Richard Hall	The Challenge of Unidentified Flying Objects
Marshall, J. S.	World of Tomorrow
Manzel, Donald H. and Lyle G. Boyd	The World of Flying Saucers
Miller, Max B.	Flying Saucers, Fact or Fiction
NICAP	UFO Evidence
New Jersey Association on Aerial Phenomena	The Shadow of the Unknown
Scully, Frank	Behind the Flying Saucers
Sullivan, Walter	We Are Not Alone
Tacker, Lawrence	Flying Saucers and the USAF
Vallee, Jacques	Anatomy of a Phenomenon
Van Tassel, George	I Rode a Flying Saucer
Von Braun, Dr. Wernher	First Men to the Moon
Wilkins, Harold T.	Flying Saucers Uncensored
Williamson, George H.	UFOs Confidential

PERIODICALS

Klass, Philip J., *Aviation Week and Space Technology*, "Plasma Theory May Explain Many UFOs," Aug. 22, 1966, p. 48.

Mallan, Lloyd, *Science and Mechanics* series, "Complete Directory of UFOs," Dec. 1966-June 1967.

Saturday Review, Sept. 3, 1966, p. 25.

Saturday Review, Oct. 1, 1966, p. 67.

www.ingramcontent.com/pod-product-compliance
Lightning Source LLC
Chambersburg PA
CBHW080456170426
43196CB00016B/2833